The Role of the
Business Analyst
De-Mystified

by

Marie-Anne Rasé

authorHOUSE®

AuthorHouse™ UK Ltd.
500 Avebury Boulevard
Central Milton Keynes, MK9 2BE
www.authorhouse.co.uk
Phone: 08001974150

First published by AuthorHouse 5/10/2010

ISBN: 978-1-4490-9624-3 (sc)

This book is printed on acid-free paper.

Dedications

To my three beautiful children: Reuben, Rebekha and Jasmine.
To the centre-point of my soul: KM

About the Author

Marie-Anne Rasé, Bsc (Hons), MBCS, CITP, has been a senior business analyst for several years. Her expertise lies in the London Market Insurance arena. She has worked in IT for over fifteen years, having held several roles throughout a typical project life cycle — from analyst/programmer to trainer to project manager. Her greatest interest has been and remains in business analysis where she is able to manifest clarity and structure from problems and chaos.

She started a career in IT in her late twenties after completing a Higher National Diploma in Computer Science followed by a degree in Business Information Systems (BIS). The latter was one of the first courses introduced in universities in the early 1990s because of many IT project failures. It was felt that one of the main factors causing projects to fail was a disjoint between the business and the IT areas. Each area was lacking an appreciation and understanding of the other. A business analyst was seen as a hybrid analyst that would understand the language, strategy, needs and issues of the business and at the same time also understand the same for the IT side.

From the time that Marie-Anne attended that degree course, she knew that business analysis would be the area she would be most interested in pursuing as a career. She certainly has made this happen.

Marie-Anne has published several Internet articles. She has recently been a participant as a panelist member in the 2009 'Women in Technology' event sponsored by the British Computer Society (BCS) and Women In Technology (W-Tech). Her journey into a successful career in IT via her studies as a mature student was used as the case study. The event brought the spotlight on the various challenges and rewards of women in technology and mainly in terms of the lack of interest from our young girls to becoming the future female technologists that are so much in need. Her passion is to motivate young girls and women alike into becoming high achievers and reaching their potential. Her motto is, "If you believe in yourself, you can achieve anything."

Read about Marie-Anne's case study and hear what she had to say as a contributor to the W-Tech 2009 Q&A panel at www.smartba.com

Drop her a line with your thoughts and comments at marie-anne@smartba.com. Share your individual experiences and give your ideas on how we can encourage women and young school leaver girls into embracing IT as a career.

Author's Acknowledgements

In no order of priority, as each had a unique and vital role in bringing this book about, I acknowledge the following people:

* My favourite Project Manager — BM. He has challenged me so much that I did not have a choice but to produce this book. Thank you for planting this seed in my mind and fertilising it to fruition.

* My three wonderful children — Reuben, Rebekha and Jasmine — for their patience and understanding as young toddlers whilst my attention and focus were directed to studying and not having fun with them. Now, mostly adults, they appreciate and recognise the rewards that demanded such sacrifices. Keep the beauty and smile in your eyes!

* My centre of gravity — the closest human being that I have met during this lifetime — KM — for bringing that ideal weight to counteract my erratic scale thereby producing a perfect balance for me to the deepest core of my cells. Thank you for being you and not giving up!

* A well-known motivational speaker — Les Brown — whose powerful words accelerated the speed of my actions to finish this book. Thank you for the inspiration.

* A last minute acknowledgement goes to Kenny Dundas who graciously brought his business analysis expert input into this book at very short notice. Thank you for your valuable feedback.

Important Note

This book is of a general nature. It is mainly based on the author's own experiences and ideas. It is not intended to be specific, customised or relevant to any particular set of individual circumstances. The readers are responsible for using their own relevant judgement to interpret the information provided for their own purposes and take appropriate decisions based on such interpretations.

Coming Soon
Business Analysis for The London Market

Table of Contents

Introduction

Business analysis was unheard of only a few decades ago. In the early 1990s, the first official courses started to appear in universities' syllabi. Why did it come about in the first place? Business analysis was created out of the myriad of IT projects going wrong in the 1980s. Until then, IT projects would solve a business problem by providing an IT system, turning manual processes into automated ones. This was very limited because the problem was looked at from an IT-solution perspective and did not factor in the business strategy, the process and the people. Therefore, many wrong solutions have used up project funding — wrong not in terms of the system not working but just wrong for what the business requires.

Thus, business analysis was born. The gap between IT and the business area would be bridged by the role of a business analyst. This person would be trained in business strategic concepts as well as information and systems technology. This person would understand how the business works and become knowledgeable in the business processes, and in turn would facilitate a solution that would fit the business needs regardless of the solution and how heavily reliant it is on information systems. The key would be that the business analyst would know enough of the problem area and the requirements to guide both the business and IT into achieving a solution fit for purpose.

Since then, business analysis has earned its rightful place in most companies. The role of the business analyst has evolved and increased in importance across the enterprise.

About this Book

The idea for this book was rooted in the frustration of the author's personal experience of the rather unbelievable expectations from others of her role as a business analyst. Ok, so the expectation did not stretch as far as her being the 'tea person' as part of her role, but it might as well have been (smiley face).

The title of 'business analyst' sounds very generic. I have yet to meet someone (outside of the corporate world) who does not ask the question, "What is a business analyst?" or, "What does a business analyst do?" straight after hearing my job title. If someone says, "I am a bus driver," or, "I am teacher," or even, "I am a project manager," people seem to have a pretty good idea of what that person does as a profession. A "business analyst," analysing businesses, what does that mean?

This book aims to clarify and de-mystify this role as well as place business analysis in its rightful context. It is a necessary read that will help you establish the concept of business analysis in terms of the following:

* where it fits within an organisation
* how it can be used best
* what a business analyst role can be and how
* how it can provide the maximum benefits to the organisation
* how it can help a project avoid some of the common pitfalls

This book uses some business and information-technology concepts and terminology. If you are unfamiliar with these, check out the glossary section at the end of the book. The Internet is also a vast sea of information on almost any topic, so feel free to search the words, ideas and concepts that you come across in this book.

Note: The material in this book is related to the general concept of business analysis — that is of a project of decent size that generally follows the main stages of a project life cycle and, therefore, benefits from the various roles at each stage. Of course, there are projects that do not follow this model; for example, a very small project may have one person fulfilling several roles of project manager, business analyst, user acceptance test analyst, trainer, etc. Larger projects or programs will have a team of business analysts working together during the analysis process and collaborating alongside many other teams for the same project.

Throughout this book, Business Analyst and Business Analysis have been abreviated to 'BA'. Likewise, Project Management has been abbreviated to 'PM'.

My Assumptions

If you have picked up this book and you are browsing or reading it, I find it logical to assume that you are either intrigued or have an interest in this topic. This can be either for yourself or for someone you have in mind. I would assume that you are looking at this book for one or many of the following reasons:

- You are thinking of becoming a BA and wish to know more about the role
- You are a PM and are wondering what a BA is meant to really be doing
- There is no such role as a BA in your company and you want to know if you are missing out on this
- You are already a BA and curiously want to see what new or interesting insight this book has to offer
- None of the above, but hey, the book looks good and 'curiosity never did kill the cat' so you wish to peek in and see how interesting the book is
- You are a cartoon fan and the cartoon characters in the book hypnotise you

Where to Go from Here

This book is structured in a logical flow so that if you have never heard of business analysis, the topic will gently be introduced to you so you are able to form a sensible mental picture. However, feel free to jump to any chapter that captures your attention. I would advise you read the whole book. I have kept it under one hundred pages just to make it an easy and light read. The cartoon characters are there to provide some pictorial humour whilst highlighting issues that do arise in real life. Business analysis is a serious matter that can make or break the success of a project.

Once you have mastered the concept and role of business analysis, be bold and curious and delve deeper into the topic. The internet is an Aladin's cave of informational treasure. Drop by my website at www. smartba.com, join the blog and grab/share some more information on the way.

Chapter 1 – Principles of Business Analysis

Would you give your money to a builder to start building a house for you without doing any groundwork first? Maybe you are not the person doing the groundwork and you will pay someone else to do it for you.

How about if you wish to get married and have your friends and loved ones with you that day — can you just wake up one morning and say, "Today I will get married"? Sure you can, but are you certain that your loved ones, especially the 'groom/bride to be,' will be available to accommodate your plans for that day?

In most of our personal enterprises, there is a level of analysis and planning that takes place. I could give you many day-to-day examples, but I will let you think about it for yourself. If you cannot even find one example in your life where you have done some form of analysis and planning, then I would like to hear from you — so contact me using the details listed at the back of this book.

The above suggests that analysis is a key stage of many initiatives. Throughout this book, I will aim to demonstrate how analysis is key to a project's success in the business world that involves information technology.

What is Business Analysis?

Wikipedia defines business analysis as "the discipline of identifying business needs and determining solutions to business problems. Solutions often include a systems development component, but may also consist of process improvement or organizational change." The person who carries out this task is called a business analyst or BA.

The International Institute of Business Analysis (IIBA) *Business Analysis Body of Knowledge* (BABoK) defines enterprise business analysis as a way for organizations to do the following:

- Identify, analyze and solve business problems and opportunities

- Determine the feasibility of a solution
- Define the solution scope and develop the business case
- Continue to assess, refine, and validate the business need and solution
- Evaluate the business benefits brought about by a solution

As illustrated in the opening paragraph of this chapter, any significant initiative that you carry out will involve some level of analysis. Business analysis, within the business context, helps define the following:

- where you are now
- where you need to be
- what you need to do to get there

Note that the 'how to get there' is not the responsibility of the business analyst. The 'how,' if involving an IT systems element, is looked at by a team of systems designers, systems analysts, systems developers and other technical experts. The BA is involved throughout to keep a check that the project is heading in the right direction, i.e., that the solution is delivering to the business needs.

Where does Business Analysis Fit?

A project generally follows a standard life cycle. There are various flavours but the stages remain fairly the same. They are as follows:

Pipeline
Initiation
Definition
Delivery
Close

Business analysis officially starts from initiation and completes at the end of definition. The business analyst's involvement, however, continues throughout the whole project cycle, interacting with other project team members. This will be illustrated further in this book.

I will briefly define the above project life cycle's stages for clarity.

Pipeline
This is a triage phase. The idea or problem is evaluated to determine its suitability for progressing as a project based on future business benefits, legal requirements, costs, etc. This is the stage where this idea or problem is given a priority and moves to the next stage of 'Initiation' based on the aforementioned criterion.

Initiation
The idea is now officially a project. Here we determine and agree on the scope of the project — what is the problem and what it is that needs to be done?

Definition
The 'what' that was identified in the initiation phase is now expanded further. The problem area should be clear enough at this stage for options to be explored. The 'what is required to get to where we need to get' is agreed and documented. Technical experts are involved and can recommend the 'how to get there' part.

Delivery
This is when the 'how' is designed and implemented. This involves sub-phases of testing and agreement that the solution is delivering to the requirements.

Close
The project's success is reviewed. A 'lesson learned' exercise takes place and feeds into improving future projects.

The core of business analysis develops from the initiation phase and matures into the definition phase. The role of the business analyst does not stop there, however. Throughout the project life cycle, the BA will assist project teams and business members in understanding the big picture rather than the tunnel vision related to each area.

The previous phases are also known under the following headings based on the popular Unified Process:

Inception
Elaboration
Construction
Transition

Other variations of project life cycle exist depending on the process or methodology being used. For example, Scrum framework would involve the phases of:

Pre-Game
Game
Post-Game

I would urge that you carry out some more research on the various processes and frameworks available for a project life cycle. This would broaden your knowledge on the various styles of managing a project.

Chapter 2 - The Business Analyst

Definition

In this section, we will explore more than one definition of a business analyst. I wish for the readers to make up their own mind of what a business analyst is from the definitions listed below.

Here is my definition illustrated as a common analogy — in one short sentence, the BA acts as a bridge between the business and the IT communities. Think of the concept of a bridge — it enables things (people, goods) that are located on one side of the bridge to transfer to the other side. If we expand this concrete concept to the idea of business analysis, we have the notion of the business analyst being a conduit from one side (the business area) to another side (information technology area). The BA as a conduit enables a two-way traffic between the two camps of IT and business.

Let's move to a more formal idea of a BA: a person who understands the business context of a problem area and translates this into a structured language that is understood by the information technology group. In reverse, the BA understands the IT context and translates this into a business language that is understood by the business crowd. Is that close to the 'bridge' analogy? I think so. If you have a better analogy, send it to me using the contact details at the back of this book and I will use it, with your permission, in my future materials.

Here are a few even more formal definitions:

The British Computer Society (BCS) proposes the following: "An internal consultancy role that has responsibility for investigating business systems, identifying options for improving business systems and bridging the needs of the business with the use of IT."

It is worth noting that, in this context, "symptoms" means whatever systematic approach the business is using, i.e. not exclusively "software systems".

Wikipedia states the following:

"The term Business Analyst (BA) is used to describe a person who practices the discipline of business analysis. A business analyst or 'BA' is responsible for analyzing the business needs of clients to help identify business problems and propose solutions. Within the systems development life cycle domain, the business analyst typically performs a liaison function between the business side of an enterprise and the providers of services to the enterprise. Common alternative titles are business analyst, systems analyst, and functional analyst, although some organizations may differentiate between these titles and corresponding responsibilities."

The International Institute of Business Analysis offers the following definition: "A business analyst works as a liaison among stakeholders in order to elicit, analyze, communicate and validate requirements for changes to business processes, policies and information systems. The business analyst understands business problems and opportunities in the context of the requirements and recommends solutions that enable the organization to achieve its goals."

In my professional view, there are two types of business analysts:

1. Those who have 'fallen' into the role either willingly or unwillingly from a different role.
2. Those who have been trained and/or mentored for the specific role of a business analyst.

The trained BA is more likely to know what needs to be done. The experience gained over many years will act almost as second nature when faced with the analysis of a project. The untrained BA is likely to inherit all tasks that do not have an owner and may take the approach of 'tell me what needs to be done and I will do it.' This is a general statement again based on my own personal experience in the world of business analysis. Before my inbox gets crowded with angry 'accidental' business analysts arguing my viewpoint, let me add this: I am sure that there exists such business analysts that have picked up the role like duck to water and are doing a swell job in that role. There... I have said my piece.

What a Business Analyst is not!

The business analyst generally is not, but may overlap, with the following:

* Strategist analyst
* Systems analyst
* Data analyst
* Solutions designer
* Architecture designer
* Systems tester
* UAT analyst

The above-mentioned roles have a space in their own right and can be performed by a different individual in certain organisations. Others will have one single individual performing all these roles. In my experience, the larger and more complex the project is, the greater the need for separate individuals doing these roles.

Skills and Role of a Business Analyst

The International Institute of Business Analysis (IIBA) *Business Analysis Body of Knowledge* (BABoK) defines enterprise business analysis as a way for organizations to do the following:

- Identify, analyze and solve business problems and opportunities
- Determine the feasibility of a solution
- Define the solution scope and develop the business case
- Continue to assess, refine, and validate the business need and solution
- Evaluate the business benefits brought about by a solution

A business analyst facilitates and drives all the above points.

The Business Analysis Body of Knowledge (BABOK) describes the common activities, tasks and deliverables of a BA as follows:

Business analysis planning and monitoring, covering: stakeholder analysis, selecting an approach to manage risks, issues, scope and requirements, deciding how to monitor and report on requirements activities, negotiating on how to manage change on the project, working with stakeholders to help them understand their requirements within the scope of a project.

Eliciting requirements: brainstorming, analysing processes and documents, running focus groups, analysing and identifying problem areas, interviewing, observing, prototyping, facilitating workshops, conducting surveys.

(Note the distinction between "eliciting requirements" which implies an activity versus "gathering requirements" which sounds more passive)

Requirements management and communication: communicating requirements, resolving conflicts, gaining formal approval, baselining and tracking requirements through to implementation.

Enterprise analysis: taking a business need, expanding and defining that need, and exploring potential solutions with the experts. It explores establishing business needs, undertaking feasibility studies and gap analysis and defining the project scope.

Requirements analysis: working with the whole project team towards defining a solution to meet the agreed requirements. It covers documenting and analysing functional requirements, quality of service requirements, identifying and modelling business domain using process diagrams and flow charts, defining the functionality using 'use cases', user-experience designs, storyboards, wireframes, user profiles and user stories, and finally verifying and validating the functional requirements.

Solution assessment and validation: assessing proposed solutions to help the stakeholders select the solution that best fits their requirements and, once selected, help the business prove that the solution meets those requirements and ultimately whether the project has met its objectives. It covers evaluating alternate solutions, conducting quality-assurance processes, supporting through implementation and post-implementation reviews.

Skills and Competencies

- Leadership
- Effective communicator
- Skilled negotiator
- Problem solving
- Business area knowledge
- Technical awareness and expertise
- Listening skills
- Analytical
- Adaptable

This chapter provides only a flavour of the role and core skills of a business analyst. As listed above, the skills required are many and varied. Amongst other factors, the nature and size of the project, the project team and the company culture will indicate the skills that would be most predominantly necessary for that particular project. The methodology used (if any) would also be an important factor in determining the role and skills of the BA and the inter-relations with other team members.

It is beyond the scope of this book to explore such topics in depth. The reader is directed to refer to mainstream BA books available in bookstores and on the Internet for more insight.

Chapter 3 – Dimensions of Business Analysis

Business analysis spans many areas of an organisation. These areas may be looked at in isolation when digging into finer details. However, the problem matter and the solution needs to be investigated across all areas. Below I explore each of these areas. They each have a weighting factor that can tip the success needle on the positive or negative side of a project.

The Four Dimensions
Business
Organisations are constantly impacted by the velocity of change in the political, economical and global landscape. As a result, the business environment is frequently changing for improvement on processes, efficiency, cost, products, services and so on. The drivers for change often originate from the business. As a given, the immediate affected business function is investigated. In addition, the surrounding business functions also need to be reviewed to ensure smooth transition to and from the affected area.

IT
Over the years, IT systems have become an integral part of the business to facilitate a competitive edge and strategic advantage. Rarely do you find core business projects that do not have an IT element within. With the merging of enterprises into a global village, IT systems are enablers of automating, streamlining and integrating business processes. IT is looked at as the solution provider (fully or partly) to most if not all enterprise projects.

Process
Can you imagine a successful company working without a process? I cannot. Maybe it might be successful only once purely through luck. Without process there is no way to hit the same target over and over again. Think about it some more — this applies to even the little things we do in life.

How important are processes? Understanding the process is key to business analysis. Not only does it tell you what is happening but it also tells you what is not happening. By looking at or actually enacting a process flow, a good analytical BA can pull out a lot of information.

"A picture tells a thousand words!" Surely you have heard this saying. Map a process on paper and let the story come out. Some details will jump at you and others you will have to find out by asking more questions or drawing the lower levels of the process.

People

An organisation can have the best systems and processes in place. They are not worth much if the people do not work the systems and the processes. So, as you see, there are interdependencies everywhere you look. Have you ever heard of a project that successfully delivered on time, on budget and to the requirements but yet was deemed a failure? It happens. Why? You need people to make it work. You need to lead and manage the changes required for the people as well as the changes to the processes and the systems. You need to be looking at the implementation of the project at the helicopter view, not at the sea-level view. The latter will only allow you to see as far as the horizon permits. Keep an eye on the people that are affected by the changes implemented by the project. These can be purely manual or automated or a combination of both. In any way, keep a close eye on the people and help them transition through the changes.

Chapter 4 - Business Analysis Today

Evolution of BA role

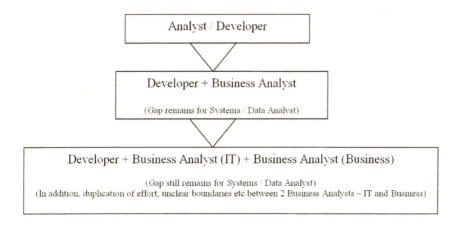

The following describes the evolution of Business Analysis:

- Stage 1 – one person is the Analyst/Programmer:
  Business analysis does not formally exist and other IT functions
  such as the analyst/programmer swallow the BA tasks. A single
  person does the business analysis, systems analysis and designs
  and produces the solution.

- Stage 2 – one person is the programmer and another one the
  business analyst:
  Business analysis is allocated its rightful place in the project
  life cycle and a business analyst role is created whereby the BA
  would look at the problem area from both the business and IT
  perspective.

  Note that the programmer now does not get involved in business
  analysis. There remains the fuzziness of who should be doing the
  systems analysis: the programmer or the business analyst?

- Stage 3 – one person is the programmer, another one is the IT

> business analyst and yet another person is the Business business analyst:
> Same as the previous step with the exception that the trend is moving towards the business analyst's role being split in two parts: the business BA and the IT BA.

Let's think about this new trend a bit more as this is the main topic of this chapter. First let me put both titles in full:

Business Business Analyst
Information Technology Business Analyst

Does that title make sense to you? It certainly does not to me! Yet, I have personally come across this, more and more at the time this book is being written. So what are the dynamics here? It is all in the split:

A business BA is 'supposed' to be an expert in the business functions and operations. Therefore, this person would do the actual 'business analysis' in the true sense of the job title, i.e., analyse how the business functions, what are the processes and operations, what is the input and output, what are the problem areas, etc.

The IT BA is 'supposed' to be an expert in IT whilst at the same time having an adequate awareness of the business areas. This person would translate the output from the business BA into an IT-solution component.

All well and good! No problem! Or is it a problem? Business analysis is only one phase of the project life cycle and the business analyst needs to analyse a problem or opportunity end to end — by this I mean including business and system processes, manual or automated. If now, one BA is looking at the business perspective and another BA is looking from the systems perspective, we end up with possibilities of 'dis-jointment' and gaps. I list below a few challenges that this trend produces:

- Where does the 'handshake' happen between the two BAs and analysis streams? By handshake I mean the smooth transferring

14

and transitioning of knowledge and deliverables between business and IT.

- Is there even a handshake as a possibility? That would assume that there is a clear-cut boundary between analysis of the business and IT areas.
- How firm is the handshake (if we assume there is a handshake happening at some stage)?
- Which BA has responsibility for what task of the analysis process?
- What about the overlaps? You can be sure that 'overlap there will be' as Yoda from Star Wars would say.
- How to ensure that both BAs do not duplicate tasks, activities, etc?
- How to ensure that knowledge acquired from users during interviews, workshops and the likes is available for both business analysts? Surely stakeholders would not appreciate having to repeat themselves for two different BAs investigating the requirements from two different angles and possibly on separate occasions.

These are but a few challenges I can think of. All is not doom and gloom though as most challenges can be overcome.

The above model can work and those challenges minimised if they are identified and mitigated before a project is started. I say this again from personal experience. The focus should be on agreeing an approach to the analysis that clearly defines the BA's roles and responsibilities. Regardless of who is responsible for what, both BAs have to be a part of the full analysis process either as the leader or the participant business analyst during all the analysis activities because knowledge that is gained during the analysis process will be relevant to both the business requirements and IT systems requirements. This is key.

A major drawback to the business BA and IT BA model is that BAs are being 'pigeon-holed.' I personally believe that BAs should be allocated analysis tasks based on their experience, skills, capabilities, drive and aspirations. It should not be only based on whether they sit behind the IT fence or the business fence on an organisational chart. A more 'friendly' and flexible model would be to have a pool of BAs, covering expertise in the business and IT functions. The BA would be allocated to a project based on a 'best fit' and availability. An additional benefit is that business analysts would have the opportunity to get involved in different types of projects thereby expanding their experience and knowledge horizon. Surely that would be a motivating factor in improving job satisfaction for your business analysts and the benefits to the enterprise would be to have even more rounded business analysts, in terms of skills.

Agile or Extreme Framework

I mention a brief note here as this topic is outside the scope of this introductory book.

Some adherents to the "Agile" or "Extreme" project framework claim that there is no need for business analysis. I think that for any project, regardless of the framework being used, there is a level of analysis that needs to be performed. This could be as simple as thinking about what needs to be done next and can be as complex as producing reams and reams of analysis artifacts. You need to be pragmatic depending on the project at hand.

Chapter 5 – Thoughts For The Future

Data and information will continue to be the lifeblood of business processes. In today's ever-changing global village, IT systems will remain a major component of business change initiatives to maintain growth and competitive edge. Enterprises will need to raise the capability and maturity of their business analysis.

Enterprises need to recognise that business analysts will need support and development to increase and improve their core competency. No doubt there is no single cause that triggers project failures. However, it is an accepted fact that the implementation of poorly defined requirements is a leading factor.

I have come across analogies of a business analyst being similar to an architect. Let's scrutinise the major differences:

An architect is regulated, a BA is not
An architect is obligated to undergo training, a BA is not
An architect needs to comply with a prescribed method, a BA does not
An architect is held liable for professional errors, a BA is not

As we can see, there are some fundamental differences between the two roles.

Business analysts perform functions that, if not fulfilled, would heavily contribute to the failure of a project. Surely, with such a responsibility, business analysts must have the skills, techniques, experience and knowledge to influence the success of the project. Business analysts should be encouraged to be trained and certified such that the practice would be standardised and professionalized.

By embracing the true value of business analysis, by demanding a high standard from your skilled business analysts, by ensuring that communication and support flow amongst the teams and by having a clear strategic directive from management, project initiatives will have a better chance of providing real value to the business.

Accept and embrace the importance that the business analyst brings towards a higher probability of project success.

This book is only about business analysis and the business analyst. At the end of the day, it is only one part (albeit a very important one) of the whole process. Delivery to high expectations and standards across the project teams cannot be ignored because after all, if a chain of the link is broken, then disaster creeps in.

Chapter 6 - BA Career – On the Move

Penetration

How do you penetrate into a BA career? We will not explore the entry by chance or necessity — i.e., you land into that role purely by accident or you are told you are now into that role so 'just get on with it!' We will be looking at a conscious decision where you wish to become a BA and a good one at that.

If there is a BA team in your organisation, then chat with an existing BA and find out what their role is. Remember that a title does not mean much; it is what you do that is important. I have known BAs whose primary day-to-day task is managing defects. I have known others who are really doing project management day in day out or who are doing data analysis. So find out what the BA does and decide whether that is what you wish to do. If available, ask to shadow a BA to get a feel of the role. Then plan your career in business analysis. Ask yourself these questions:

- What skills do I already have that would be required as a BA?
- Is there any books/reading material that I can read to help me understand business analysis?
- What training will I need?
- What training is currently available that I can attend?
- Will my company pay for my training?
- Is there a process where I can easily ask for a transfer into the BA team?

If a BA team does not exist in your organisation or if you do not belong to an organisation, then still ask yourself the first four questions above. Create a career plan. This is key. This helps you to see clearly your objective and what is required for you to reach that goal. The plan also acts as a measuring mechanism so that you can keep track of the distance between you and your goal.

The next step is to then take action and take the steps that will draw your goal nearer to you.

Let me share my story very quickly with you: I was a medical PA 'temping' for an agency. This means that the agency would place me in organisations that require temporary staff to cover for sickness, maternity leave, etc. One day I decided I was going to do something else. I went to university as a mature student, completed a Higher National Diploma in Computer Science followed by a degree in Business Information Systems. I then launched into an IT career.

This is a fifteen-odd-year story condensed in only a few lines. The path has not always been rosy. If you are interested in finding out about my journey, come and see my case study at www.smartba.com

You can make your BA career happen. My story proves this.

There are many entry points and my aim is that those ideas I have presented here will springboard others that are most relevant to your situation.

Where To?

So you are a proficient business analyst. You have either reached your peak or would like to explore new skies. What is out there for you from where you currently are?

Once you have peaked as a senior business analyst, there may be a sense of a lack of options for a career progression in the same field. The most obvious move is to become a BA team lead or a BA lead. What is the difference?

BA team lead – you become the manager of a team of BAs. Usually you are their reporting line and your tasks become more administrative in terms of managing the team rather than doing practical BA work. Depending on how many business analysts you manage, the time spent on analysis tasks will decrease.

Lead BA – here your practical tasks as a BA remain prominent. In addition, you are co-ordinating the allocation and delivery of tasks for a group of BAs who are working on the same project(s) as you. This role is usually present on larger projects or programmes. You hold the lead role as long as the project is in place. You are not the line manager of those business analysts but you manage and coordinate their tasks and deliverables as well as your own.

These two roles may not be readily present as an opportunity for you in your company due to its structure and size. So where do you go then? Many BAs move into a project management role. Although some BA skills overlap those of a PM, my view is that the core skill-set is quite different. Being an excellent BA does not automatically make you an excellent PM and vice versa. I would advise that some formal PM training and mentoring would be highly recommended if you decide to become a PM rather than a BA. As any other role, you can either be good at it or bad at it — the choice is yours.

In the current time, a few more options are available as a career progression that keeps in line with the BA practice and will still use all your BA skills and experience. Let's have a look at what they are:

As mentioned earlier, the BA role is sometimes split between the business BA and the IT BA (oh! how I dislike this title!). Let's move on. If your background has been predominantly in the IT area, then an option is to become a BA on the business side. In that role, you will use all your existing skills and experience. Your progression will be to build your expertise in the business functions hence complementing your existing expertise. The reverse is also possible. If your business analysis background is from the business area, then you would switch to the IT area. Both scenarios will warrant some training and will provide new challenges and opportunities in improving your BA skills.

Another opportunity is to specialise into a specific area as business analyst. For example, specialising in Business Process Modelling (BPM). BPM would become your niche that would warrant a new learning curve and formal training that would give you the credibility and professionalism you want. An example would be to train as a Lean Six Sigma process specialist.

These are just ideas that you can explore. Do not believe that you can only progress horizontally once you have reached your career peak as a BA. Vertical progression is still possible. Any decision that you make to expand your current skills and expertise will require some form of commitment if you wish to be very good at it. The question is, "Are you prepared to do whatever it takes?" If not, then you can still remain a BA as you are currently. That too is an option.

Show Me The Money!

Money you shall see indeed!

In today's terms (at the time of print), employers are offering circa £65k for a highly skilled, specialised and experienced BA for a permanent position in London UK. Contracting rates can bring circa £450/day again for such a BA. 'Experienced' usually mean that a person should have at least ten years' experience in the role.

Ask yourself these questions:

- How experienced are you or how experienced do you wish to be?
- How skilled are you or how skilled do you wish to be?
- What area do you specialise in or would you like to specialise in?
- How confident are you?
- What training have you had in the past and will you require in the future?

The answers to the above will give you more or less bargaining power to your earning potential.

Remember though that salaries will fluctuate depending on the sector you are working in. Thus, financial services will generally offer more money than public sector or non-profit organisations.

My point is that a career as a business analyst can and will be well paid if your skills and expertise tick the right boxes.

This information is just to give you an idea. Do your own research on job advertisement sites and materials and you will find the range that is relevant to the current time.

Chapter 7 - Ten Best Practices in Business Analysis

There are many dynamics at play in making a project a success. Here I will point out some of the main ones that will improve the chances of success. The scope of this book does not include all the factors that can surface. The guidelines below can surely be used as a starting point.

Analysis is Key

Too many times, project sponsors and decision makers do not see the need for rigorous analysis in a project. "We have a problem, let's just fix it." If you have a leaking water pipe, do you just fix the pipe where you see the water dripping or do you trace the leak to its source first and then fix the source of the leak? Do you then go even further to finding out why the leak started in the first place and try to prevent it happening again? Then do you also fix the damaged surrounding areas around the leaking point?

Likewise for a project. I think you get the point. Projects are usually so much short of time and money that Post Implementation Reviews (PIR) are conveniently forgotten. This is a stage where the project is reviewed for good points and bad points. The lessons learned can be used as valuable 'not to be repeated again' for future projects. In all these years I have worked in IT, I have yet to see a project that first does a proper PIR and second that actually uses the result to improve future projects. More often than not, the teams are busy starting on the next project and thus that valuable PIR phase is brushed aside.

So analysis is key. How much analysis is required is dependent on the type of project — its size and complexity. If analysis is not required, this also needs to be documented with the reasons as part of the project documentation, i.e. it should be a deliberate decision and the risks (if any) associated with not doing analysis should be documented.

Predictive Clairvoyance

No, you do not need a crystal ball! You do need, however, to have an eye for details as well as an eye for the bigger picture. It is very tempting and easy when looking at a problem area to focus only on that specific area. This may not even be the problem but only a symptom to the problem. Therefore, look outside the box. Just like the leaking pipe illustration, trace the problem to its source.

Think ahead of the problem area. Think about impact, interferences, dependencies and inter-dependencies. Think about it as if you were visiting your General Practitioner (GP). Usually you would only go there when you have a problem. Your GP will find out historical information that would help diagnose the cause of the problem. Then the GP will prescribe something (a potential solution), inform you of what to expect and your next step if the expected result does not materialise. Let's relate that to our subject — business analysis. You will be informed of a problem, you will look at all the history around the problem, you will identify the cause, explore potential solutions by predicting end results and have a measurement for success mechanism in place for the end of the project. Of course, this illustration is the happy and direct route but we all know it does not happen that way. The reality is more in line with more problems being put on the table, businesses needing change, expectations for solutions getting stretched to the point of becoming unreal and measuring success remaining within the ether of the plan.

You can see why you need predictive clairvoyance. Plan, plan and plan some more; and then review the plan and re-plan again. So what if you have to re-work the plan several times? At least you have a reference point to measure how close or far you are from the goal.

Watch the gaps. Looking at the bigger picture and then moving on into the more detailed levels one by one in a consistent manner will reward you with identifying gaps. Until you find out about these gaps, you cannot mitigate them. You certainly do not wish to find gaps when the project is coming to a close and the solution has been delivered. This would result in a labour- and money-intensive exercise to rectify.

Team Synergy

Picture a medical surgery — as a minimum you will have an anesthetist, a surgery nurse and a surgeon. Each knows what needs to be done. Each appreciates that the success of the surgery depends on each person performing the individual task properly and on time. Would you lie down on the operating table knowing that the surgeon and anesthetist have swapped places to perform the surgery? You would be insane to! So what is my point?

Just that! Team work in any project is key. Let each team member do her or his job properly. Support each other and provide assistance where required and 'let business analysts get on with it' and likewise for the other team members.

BAs should know the people on their team, including their business users, to understand their working style and adapt to it. Once the core analysis of the project is over, they are the reference point for many issues, queries and guidance.

Team working is a subject in its own right and is beyond the scope of this book.

Mind Your Expectations

Know what is expected of you and clearly communicate what you expect from others. From the start of the project, the business analyst needs to agree and understand the roles and ownership of each team member. The BA is the hub of the project being the closest to understand what the business needs are. Be clear on the following:

- what the analysis stage consists of
- what are the input into the analysis and who are the providers of the input
- what will be the output of the analysis and who will be the recipients of the output
- what support the BA will need to provide and to whom
- what support the BA will need and from whom

More and more nowadays, the traditional BA role is being split into two camps: the IT camp and the business camp. In such organisations, the waters get muddy and ownership and responsibility of BAs overlap or have holes and become confused. Result? Gaps in analysis due to one camp thinking that the other will be responsible for such or such activity or duplication of efforts and tasks. Hence, minding your expectations and that of others is key.

Challenge others, ask questions and make sure you are clear on the answers — who does what, when and how? For deliverables, ensure that the format and medium are agreed, especially where no templates or existing reference documentation exist.

What is Your Language?

A key part of business analysis is documentation. Remember you are the navigator who identifies 'where the business is,' 'where the business wants to be,' 'what is required to get the business from here to there.' It does not help the other team members if all this information is secured safely in your brain. Documentation is key! What you document and how you document need to be carefully planned. You can do all the analysis you want, but if you cannot communicate this at the right level and to the right audience, your end result solution is likely to be a deviation from what you had in mind.

If your organisation uses existing templates and standards, then it is easier for the BA to document. There are many benefits to having standard templates that are used for projects. Nowadays, the Unified Modelling Language (UML) is seen as the dominant favourite for the business analysis community. At its highest definition, UML is a set of graphical notations used to describe and design a system.

Whatever documentation style used, whatever methodology used, whatever tool used, the key is to ensure consistency, accuracy, traceability and completeness in all documentation. Ensure the business analyst is trained to the organisation's standards.

Complete the Puzzle

Look at the whole picture. If bits are missing, then find out what is missing. The business analyst must, let me repeat, <u>must</u> take the whole picture in context, not just the problem area. This goes as far as the input and output to the whole picture. How else can one ascertain the impact, dependencies and gaps that may be present or may be introduced once the problem is resolved?

The key to business analysis is looking at the full picture from an end-to-end perspective. This should cover all manual processes as well as system processes from beginning to end within the business area in scope.

The Change Cycle

So you have your perfect systems solution in place, you have ensured all the systems processes work, you have measured your systems delivery against your requirements and green ticks are in order. You feel very proud of yourself…yet there is a nagging feeling at the base of your brain!

The people! People and change mix as well as oil and water. You forgot about change management! People are individuals of habit. Our comfort zone is a highly magnetised area and it takes great effort and courage to put a foot outside of that zone. What am I saying? As you are preparing your system and process solutions you also need to prepare your people for the changes that are going to intrude their habitual routine.

Look at the Change Cycle Wheel below – this is one interpretation of many other change cycle wheels. Even for the smallest of changes in our personal or professional lives, we go through a form of change cycle. Unless it is a big change, we fail to notice that we are going through these stages. Do not let change control off your radar even when your attention is busy elsewhere.

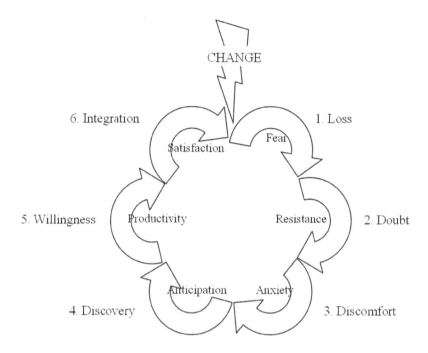

What Is in your ToolBox?

Templates, standards, methodology and tools are useful elements of business analysis. A builder would have a method of building a house (foundation first, etc.), would have all the tools necessary to build the house and would have some building regulations to adhere to. In a similar way, an enterprise that has certain guidelines, standards, methodology and tools in place for conducting business analysis would greatly enhance the tasks of the business analyst. In addition, these may be used to improve consistency, completeness and quality of business analysis across projects in the organisation.

Draw the Line

What does drawing a line provide? A reference point to measure against in the future. It is vitally important to get approval of all project documentation during analysis — this is referred to as a 'baseline.' First, it aids as a confirmation of the accuracy of the documentation.

Secondly, it confirms the scope of the project and enables the business analyst to manage any new items that creep into the project. Thirdly, it is used as a blueprint of what the problem or opportunity is and what needs to be delivered.

Let the Show Begin

Imagine you are at the circus! You have the clowns, the trapeze swinger, the lion tamer, the acrobats, etc. You would not expect the clown to be doing the trapeze swinger's act or vice versa, would you? It might be funny at first but soon you will feel that you have been short changed. So, why would you want your BA to be your systems tester or your solutions designer or your architecture designer? Let the experts do the job they are expert at. If there are such experts, do not try and assume their role or make decisions for them. Teamwork is key as is working to the same objective. Anything less would bring chaos and lack of clarity resulting in higher probability of failure to the project.

Chapter 8 - Twelve Pitfalls to Avoid in Business Analysis

What Is the Problem?

Take the time to find out what the problem is. Many times, problems are explained as symptoms. Sometimes you do not even get to hear about the problem; you just hear that a solution is required! A business analyst should get used to ask the key questions: 'who', 'why,' 'what,' 'where,' 'when,' 'how' and 'how many.' It is by asking questions that the real problems can be teased out of people's explanations.

Be clear of the problem. If you get this wrong, your end result surely will be wrong too.

Jumping to Solutions

When you go to see a doctor to report a medical problem, do you also tell her or him what you need as medication for that problem? I have witnessed many times, in workshops, clients telling their business problem and then straight away identifying the solution that is required for them. Notice that usually business users will be seeing and identifying the problem at their level only, i.e. tunnel vision of the problem. So the solution they mention will be to resolve only that problem regardless of what impact it would have elsewhere in the process. Both IT and business are guilty of this tendency. It is very, very tempting and great control is required to keep looking at a problem at the correct level.

Analysis Paralysis

Is there such thing as too much analysis or wrong analysis? Of course. Spending too much time analysing too much in details can be a blockage for a successful project. Constant scope creeps or badly managed scope can stretch the analysis stage on and on and on.

If the scope of the project is unclear, clarify it as much as you can and get it agreed and baselined. Once this is done, it is much easier to manage new or changed scope. Of course, the business analyst needs the support from the project manager and the project sponsor in this task. Remember, team synergy is key.

Analysis Denial

BAs sometimes struggle to get the analysis spot required in the project life cycle. Often, project sponsors or senior project team members do not recognise or accept the time required and the importance of conducting a thorough analysis as improving the success probability of the project. In projects where analysis is readily accepted as a necessary phase, the time required for the analysis may be seen more as increasing the project budget and time. The resulting effect though would be that less re-work (that would be expensive) would be required at a later stage of the project. The later a problem is found, the more expensive it is to rectify.

Proper analysis will help identify the scope of the project, define a measurement for success and even aid in decision making of "go/no-go" for the project.

Size Does Matter

Of course it does! You try to fit one analysis approach to different projects of different sizes and complexity and then do a post-mortem review. Yes, it will indeed be a 'post-mortem' review!

For each project, the business analyst should adjust the analysis approach and terms of engagement as necessary. For small projects, some documentation can be skipped, for others all documentation is a must. Some projects have keen and informed stakeholders and your engagement with them would be more on the smooth side. Others, well you work it out.

Each project is different so evaluate it accordingly. You would usually wear a different attire style depending on the occasion, right? Therefore, use a different attire for each project you do as dictated by that project.

Single Level — Single Facet

Do not allow yourself to be led by the narrations of what the problem is. You will certainly be having conversations, interviews or workshops to find out what the problem area is. People have tendencies only to see what is most related to themselves.

Business analysis conducted at only one level of the business is prone to have gaps and inaccuracies. At whatever level you start your analysis, be sure to investigate the higher and lower levels of the problem areas. Joiners, leavers, adjacent, parallel and duplicate areas/processes must be looked at too. The solution will be based on the requirements, the requirements are based on understanding the problem area and all its dependencies and inter-relations. These are related as a chain — get something wrong at the beginning of the chain and surely the end of the chain will reflect that mistake at a potentially high cost.

Take care to look at the problem area from all angles and at all levels.

The Creepy Crawlies

What has a tendency to creep into a project and is guilty of increasing cost and time? That's the crawlies scope creeps that, out of the blue, make their way into the project.

Managing scope is an art that even the most skilled BA amongst us need to be vigilant about. Caught off guard, the scope increases and everything from the estimated time to the estimated cost can go offline.

Where is the Action?

If you wish to buy some plants, do you go to the car manager to ask for advice or do you go to the gardener? Of course, it will be the latter because he knows about plants — that's his day-to-day life.

Be where the action is! When doing a project's analysis, track those people who are doing the day-to-day job as your target. Senior managers are there to make decisions, usually high-level decisions. When it comes to finding the problems and understanding how things work, you need to be where the action is.

Ignore the Soft Stuff

Processes are great. Systems are brilliant. However, you need people to make them work. Keep them involved in the project. Ignore the people and, no matter how incredible your processes and systems are, your project will be a failure if the people do not use them as they need to.

Well, one option is to sack the people who resist…this is a topic that we will not dig into at this stage…

Forget the Tape Measure

Of course you cannot forget the tape measure! Well, a tape measure won't do anything for your project. However, a form of measuring your project's progress must exist. How else will you know whether the project hit the target or missed?

Being able to measure goes back to requirements again. It is amazing how requirements seem to be the 'centre of the world' of a project!

To be able to measure success, you need to be able to trace the business benefits back through the requirements to the original needs/drivers. This implies that each requirement would have been defined with a view of achieving a business benefit. If there is no business benefit, there is no requirement. Requirement defined to business benefit also drives the prioritisation process of requirements — all key to delivering the solution whilst managing the time and budget.

No Analysis Please

I have aimed to demonstrate that 'no analysis' is bad for the health of a project. The balance is the right analysis by a skillful business analyst, supported by the right teams from IT and the business.

So a project without analysis is a ship heading straight into an iceberg. Actually, let me correct this — the ship is already on the iceberg!

The earlier in a project that a bad choice is made — as an example agreeing on a solution without looking at the whole problem — the greater the damage in terms of how much re-work and maintenance may be required. This is compounded by the size and complexity of the project.

Do not miss the opportunity to prevent this disaster.

Role Model Wanted

A formula for disaster is engaging in a project without a clear idea of each other's role and responsibilities. I find it very confusing having someone expecting me to do something and not having a clue what I am supposed to do. Have you been there?

The role of a business analyst easily overlaps surrounding roles such as the project manager, the systems analyst, the systems designer... and the list goes on. Why is this? Think of the business analysts as the hub of a network — most of the actions and knowledge goes through them. Many decisions and clarification are passed via them. The BA's deliverables feed into downstream teams and they have an appreciation of the technological language used further down the line to realise the solution. Upstream, the BAs know about the risks/issues/scope of the project and frequently will be approached in the absence of the project manager.

However, as the medical surgery image illustrates, each role has its unique task to perform. Having one person doing many roles in a fairly sized project is increasing the chances of a dilution of the quality of the tasks performed.

Therefore, know your role, what is expected from you and that of everyone else on the team.

-- END --

Coming Soon

Business Analysis for The London Market

Appendix A: Glossary

Agile   Agile refers to software development methodologies that are based on iterative development.

BA   Business Analysis or
Business Analyst

BABoK   Business Analysis Body of Knowledge

BCS   British Computer Society

BIS   Business Information Systems

Blog   Blog is a contraction of the term "web log". There are many types of blogs – a common one being a website that provides commentary or news on a particular subject.

BPM   Business Processing Modeling

CITP   Chartered Information Technology Professional

Extreme Framework   Extreme is development methodology based on the Agile framework (see above). It advocates frequent implementations in short iterative cycles.

GP   General Practitioner

IIBA   International Institute of Business Analysis

IT   Information Technology

Lean Six Sigma   Lean Six Sigma is a strategy to improve the quality of processes.

MBCS   Member of British Computer Society

PIR   Post Implementation Review

PM   Project Manager or Project Management

Project   Any change that is brought about to an organisation's processes, products or services is known as a project.

Project Life Cycle   The stages that a typical business project goes through. An example of a project life cycle is Initiation, Definition, Delivery, Close.

Q&A   Questions & Answers

SCRUM	Scrum is an incremental development framework that is structured around iterations of work.
UAT	User Acceptance Testing
UML	Unified Modelling Language
Unified Process	A framework that can be customised to specific organizations or projects that describes the processes involved.
W-Tech	Women in Technology